TROPICAL TREES

of the

PACIFIC

Text and Color Photography

by

Dorothy and Bob Hargreaves

Published by
ROSS-HARGREAVES
A Division of L. & M. Equipment Co. Inc.
P.O. Box 11897 LAHAINA, HAWAII, 96761 U.S.A.

AFRICAN TULIP TREE, Flame of the Forest, Fire Tree, Fountain Tree

· French: Immortel Étranger, Baton de Sorcier, Tulipier du Gabun

Spathodea campanulata Beauvois

Discovered in 1787 by Palisot Beauvois in Africa, this ever-blooming fiery red tree can now be spotted throughout the tropical world. There are many lovely trees in New Caledonia, and Fiji. The name Fountain Tree comes from the water contained in the unopened buds (see picture right) which squirt water when squeezed or pierced.

2

ALMOND TREE,
Tropical Almond,
Indian Almond
Malabar Almond

· Ceylon: Kadoru · Fiji: Tavola, Tavola Lato, Tivi · French: Badamier, Amamdier des Indes · Hawaii: Kamani · Malaysia: Ketapang, Lintak · Philippines: Talisay · Singapore: Ketapang Almond · Tahiti: Tamanu, Autera'a · Tonga: Fetau

Terminalia catappa

As this 30' tree withstands salt spray, it is seen extensively throughout the tropics in S. E. Asia and the South Pacific Islands. It can be recognized by its horizontal branches which grow in wide spreading circles at different levels on the trunk, by its large leathery leaves which turn red before they fall, suggestive of Autumn leaves, and by its flat almond shaped edible fruits. There is a variety in Ceylon which produces the very good "Okari Nuts" (*T. okari*). Probably one of the finest of tropical nuts.

AMHERSTIA, Pride of Burma, Orchid Flower, Rose of the Mountains

· Java, Burma· Toha, Thawka

Amherstia nobilis

This "Queen of the Flowering Trees" is said to be the finest flowering tree of the tropics. It is named after Lady Amherst, wife of a Governor of Burma. The original tree was discovered by Dr. Wallich, Director of the Botanic Gardens, Calcutta, in the Teak forests of Burma. The oldest specie in Ceylon was planted in 1860 in the Royal Botanic Gardens at Peradeniya. There are many lovely trees there now. The beautiful vermillion colored blossoms which covered the ground were a daily offering to Buddha. These 8″ by 4″ orchid-like flowers dangle from a central stem on each branch and truly look like a huge bunch of orchids. They are superb.

4

ASOKA, Sorrowless Tree

• Ceylon: Diyarambala, Diya-ratmal
• Malaysia: Gapis, Golak

Saraca indica

A 30′ evergreen tree of India and Malaysia with large, sweet-scented yellow and orange-red flower heads which look something like the Ixora (*Hargreaves*, "Tropical Blossoms of the Pacific"). The tree is sacred to Hindus and Buddhists because Buddha is believed to have been born under it. Like the Amherstia, a relative, the dark green leaves have drooping pointed leaflets that hang like a tassel several days before stiffening and straightening. They have large purplish black 10″ pods with four to eight oblong seeds that the monkeys and squirrels eat. The Thai people eat the flowers and leaves of one species. Can be seen in Bogor Botanic Gardens in Indonesia; Royal Botanic Gardens, Ceylon; Botanic Gardens, Singapore; Hawaii and the Waterfall Gardens, Penang, Malaysia.

5

BANANA Musa

- Bali: Biju · Fiji: Vudi, Vudi ni Vavalagi
- Guam: Aga · Hawaii: Mai'a · Loyalty
  Group: Wa-wi-wi · New Guinea: Olomp
  bomp · Philippines: Saguing, Pisang
- Rarotonga: Meiki · Samoa, Tahiti: Fa'i,
  Fei · Thailand: Klue · Tonga: Siaine

There are over 300 edible forms of bananas.
The trees are used for food, roofs, cattle feed,
clothing, medicine, dye, alcohol, vinegar,
packing material, etc. Once the tree has
fruited it dies, but a new plant has sprung
up beside it.

FLOWERING BANANA
M. rubra syn. M. orna

6

BANYAN—800 species ficus (Fig)

· Bali: Beringin · Philippines: Baleté (Tag.) · Fiji: Bakawai, Nunu
· Samoa: Mati · Guam: Nunu · Tonga: Ovava Kahi

F. benghalensis

INDIAN BANYAN, Vada Tree (above) is named for the Hindu traders named Banyans. It is a huge evergreen sacred to the Hindus. The aerial roots grow earthward from horizontal branches supporting the tree, so that it covers large areas. In India, its native land, one measured 2000′. It was 85′ high.

BANYAN (continued)
BO TREE, Peepul Tree, Sacred Tree

- Ceylon: Bodl, Pipal, Aswatha, Aracha • Malaysia: Bodl, Pipal, Peepul, Ara • Thailand: Po Tree • Burma: Bawdi Nyaung

Ficus religiosa

This smooth grey trunked tree was first brought from India to Anuradhapura, Ceylon in 288 B.C. It is the oldest historical tree known. The parent of all still flourishes in Ceylon. King Edward VII planted a Bo Tree in the Royal Botanic Gardens, Paradeniya, Ceylon in 1875. They are planted beside each Buddhist temple, because Buddha meditated under a Bo tree for six years.

BAOBAB, Dead Rat Tree, Judas Bag, Monkey-Bread Tree, Bottle Tree

· Ceylon: Tebeldi, Purunku

Adansonia digitata

From East Africa, this is one of the largest (around) and longest lived trees in the world, ranking with the Sequoias. It has a thick trunk that grows up to 100′ around. Often this trunk becomes hollow and can hold up to 250 gals. of water. It is one of Africa's most useful trees. The leaves, bark, and fruits are used for food, cordage, paper, thread, homes, medicine, and cloth. The 6″ pendulous white flowers and dark green foliage are handsome. The 12″ by 4″ long oval fruit hanging from a long stalk contains about 30 seeds, thus suggesting the name "Judas Bag". Trees can be seen in Jaffric, Ceylon; Penang, Malaysia; New Caledonia; Australia.

9

BREADFRUIT

- Bali: Sukun Timbul · Ceylon: Del · Fiji: Uto · Guam: Lemai · Hawaii: 'Ulu
- Samoa: 'Ulu · Malaysia: Sukun · New Caledonia: Arbre Végétal · Philippines: Rimas (Tag.) · Tahiti: Uru Thailand: Sa-ke · Tonga: Mei

Artocarpus incisus

A 30′ to 60′ lovely tropical Malaysian tree with exotic split leaves ranging from 1′ to 3′ long. Fruits weight up to 10 lbs. The wood was used for canoes, bark for tapa, sap to fill in the seams of canoes and tapa. One or two breadfruit trees provides food enough for a year for a family. The fruit is high in carbohydrates, and is a source of Vitamin A. B. and C.

10

BROWNEA, Rose of the Mountains, Scarlet Brownea, Rose of Venezuela

Brownea grandiceps

This handsome 40′ tree, which is a native of Venezuela, was introduced into Ceylon in 1870. The bright red flower heads that are often 8″ across, bloom continuously. The long drooping leaves protect these blossoms from the bright rays of the tropical sun during the day, but at night, the leaves lift and expose the blossoms to the dew. Found in the Royal Botanic Gardens in Ceylon, Bogor Gardens in Indonesia, Waterfall Gardens, Penang, and the Foster Gardens in Hawaii. Also found in Fiji, the Philippines, and Malaysia.

CANDLE NUT TREE, Indian Walnut

· Ceylon: Kekuna, Tel Kekuna · Fiji: Lauci, Tutui, Sikeci, Qereqere, Toto · French: Bankul · Hawaii: Kukui · Malaysia: Buah Keras, Kemiri, Kembiri · Philippines: Lumbang Bato · Samoa: Pu'a, Lama · Tahiti, Marquesas, Tonga: Tuitui

Aleurites moluccana

Brought to the islands by early Polynesians who used the oil for stone lamps, kernels strung on coconut midribs for candles, and husks and the roots for tapa cloth dye. They are from the Molucca Islands, Malaysia, and the S. Pacific. Many of these islanders make leis like the ones below out of the nuts. They are also used for varnish, medicine, relish, and fertilizer. There is a tree (*A. fordii*) that produces Tung Oil. These have been planted by the Forestry Dept. in Hong Kong.

12

CANNONBALL TREE

• Ceylon: Dagoba Sal • French: Arbre a Bombes, Boulet de Canon, Abricot de Singe

Couroupita guianensis

This huge tree is a novelty wherever it grows. The 3″ to 5″ waxy sweet smelling flowers push right out of the heavy bark and have no connection with the foliage at the top of the tree. These are followed by hard shelled cannonball like fruits 6″ to 8″ that are used as calabashes. Their pulp has an unpleasant odor. Can be seen in the Royal Botanic Gardens in Ceylon, the Public Gardens of Kuala Lumpur, the Botanic Gardens of Singapore and Penang, Malaysia, Foster Gardens of Honolulu.

CASSIA, Golden Shower, Pudding Pipe Tree, Indian Laburnum, Purging Cassia

- Ceylon: Ehela, Tiruk-kontai
- Philippines: Kañapistola (Tag.)

Cassia fistula

This Cassia has large clusters of bright yellow blossoms which cascade down from its branches making the tree look as if it were always in the sunlight. The flowers have five petals with a spike-like blossom. Long curving pistil and stamens project from the center of the flower. The pistil develops into the round black pod which grows up to three feet in length. The pod gives the tree the name of "Pudding Pipe" in India where a cathartic is made from its sticky brown pulp and sometimes added to tobacco. The flowers are used as temple offerings, the bark for tanning and native medicine.

PINK CASSIA, Pink and White Shower Tree

· Philippines: Apostola (Tag.), Anchoan

Cassia javanica

Another of the Legume Family is this small apple-blossom like tree from Java. Its branches are completely surrounded by masses of unevenly tinted pink flowers. Each petal is pale pink or white with deeper pink veinings, giving a variegated effect.

COVER: **Rainbow Shower Tree**

Cassia fistula × Cassia javanica

One of the loveliest Cassia hybrids is the Official Tree of Honolulu. It is a cross between the Pink and Golden Shower. They bloom for a long time in all their bright two-toned, breathtaking, fluffy array.

15

CREPE MYRTLE, Queen of Flowers, Rose of India

· Bali: Tangi · Singapore: Bungor · Ceylon: Murutha · Thailand: Tha-bag,
· Malaysia: Bungo, B. Raya · Philippines: Banaba · Burma: Pyinma

Lagerstroemia speciosa

Another one of the most brilliant floral displays is a native of India where it is called "Jarool". Used decoratively and as timber because it is tough and hardy. It is a favorite in S. E. Asia, Australia, the Philippines, Malaysia, Singapore, Guam, Ceylon, Thailand, Vietnam and Cambodia, as well as many other places. Its

2″ to 3″ rose-like blossoms vary from mauve to various shades of pink, have six fluffy petals, and oval fruit that is also quite pretty. The tree is useful for medicinal purposes—the leaves for diabetics as insulin, the bark, roots, fruit for astringent. (They are 17% tannin). The seeds are narcotic.

DOMBEYA, Hydrangea Tree, Pink Ball, Mexican Rose, African Mallow

• Phillipines: Anapola

Dombeya wallichii

Showy, drooping, pink scentless balls, like Christmas ornaments, bloom at Christmas time among the dense foliage of this striking tree. Each 1″ flower has five petals. Many of these flowers are crowded into the showy heads that form the lovely clusters. They continue to hang on their long hairy stalks long after the color has faded and they have turned to brown. They are occasionally cultivated in Guam, Malaysia, New Caledonia, and the Philippines. Can be seen in Foster Gardens in Hawaii.

DRAGON TREE, Dracaena, Dragon's Blood

Dracaena draco

This is said to be the oldest vegetable inhabitant on earth. Dracaena is Greek for dragon. The sap of this tree from the Canary Islands is supposed to resemble dragon's blood. It is used as a coloring for varnish, and was used medicinally in ancient times. The excellent tree above can be seen in the Botanic Gardens in Sidney, Australia.

DURIAN, Civet Fruit

· Bali: Duren · Thailand: Turian · Burma: Duyin

Durio zibethinus

Another large tree famous for its spinelike fruit is said to be one of the most popular trees of S. E. Asia. It is indigenous to Malaysia, where it is not only eaten raw, but preserved in salt to eat the year around. The seeds, too, can be roasted. The leaves of the tree are 3″ to 7″ long, and 2″ white flowers cluster on the branches where the large 10 lb. fruits develop. Their pulp is very delicious, a creamy white custard, but the fruit has a very disagreeable smell—the botanical name comes from the Italian word "zibetto" or civet—meaning strong smelling. The people say, "The fruit is heaven, but the smell is hell". Royal Botanic Gardens in Ceylon, Philippines (University), Thailand, Indonesia, and Malaysia.

EBONY

· Bali: Eban · Malaysia: Ka-
yu, Sihangus, Bu Buez

Diospyros ebenum

The Ebony Family includes
about 300 species of trees with
hard dark wood, some yielding
edible fruits, some valuable
timber. The best known cabinet
wood is the heart-wood of Ma-
cassar Ebony. This hard, fine
grained wood is black with
brown stripes and takes a
handsome high polish. The
people of Bali make beautiful
carvings from Ebony. Trees
found in Ceylon, Malaysia,
Indonesia.

GEISSOIS

• Fiji: Vuru, Vuga, Voga

Geissois pruinosa

This species of tree is found in New Caledonia. The blossoms stand out from the old wood of the tree like a big brush for bottles. The flower is similar to the Callistemon (Bottle Brush) "Hawaii Blossoms" *Hargreaves*, Page 40. Other species can be found on Fiji and other S. Pacific islands and Australia.

HORSE-RADISH TREE,
Drumstick Tree, Ben Tree

• Bali: Kelor, Pala • Ceylon: Murunga, Mur-
unga-Kai • Guam: Calamondin • Philippines: Calamunggay Tree, Marunggay,
Marong-Gay • Samoa: Lopa

Moringa oleifera

Fern like leaves and 1″ yellowish white flowers,
blooming continuously, characterize this useful
tree that looks like a Legume. It has many 18″
beanlike pods. The young pods are used as a
vegetable. The leaves, shoots, flowers are also
used. Corky bark used for gum; seeds fried or
roasted taste like peanuts; also used in curries.
Roots, when grated, taste like horse-radish—
thus the name. "Oil of Ben" is extracted from
the seeds. The different parts of the tree are
used for salads, soup, perfume, lubrication,
and medicine. Easily propogated from cuttings.
This is much cultivated in Bali, the Philippines,
Malaysia, Indonesia. Most people in these
countries have at least one or more trees in
their yards. Some also in New Caledonia, Fiji,
Hawaii, Guam, Ceylon, and Samoa.

IFIL TREE, Ipil Tree

- Guam: Ithia
- Samoa: Ifilele

Intsia bijuga

This is the Official Tree of Guam. The picture was taken in a small park with reconstructed "Latte Mepo" in the background. (These are ancient monuments of the inhabitants of Guam found on Guam and other islands of the Marianas.) The tree is a small tree which will grow near the beach in sandy soil. Each flower has a downy calyx, one white or pink petal and three stamens. The pod is thick and leathery 4″ to 12″ long, and about 2″ thick. Wood is useful for interior work.

23

INDIAN CORAL TREE, Tiger's Claw, Crabclaw

· Ceylon: Murunga-mara, Erabadu, Mullu-murukku · Fiji: Drala, Rara, Rarawai · French: Follie de Jeune · Guam: Gabgab, Gapyap · Malaysia, Phillippines: Dapdap, Dadap · Rarotonga, Samoa, Niue: Gatae, Ngatae · Tahiti: 'atal · Burma: Kathit

Erythrina indica syn.
E. variegata var. *orientalis*

This large, spreading deciduous tree bursts into pointed red pea-shaped blossoms in almost all tropical areas. These long spikes jut out of woody stems on the ends of the branches. The individual flowers break out of the split side of a pointed calyx with one flower petal much larger than the others, giving the effect of a pointed claw or feline toenail. The seed pods, too, resemble pointed claws (see right above). The tree is a Legume from Asia. Natives take the bark and seeds to stupefy fish so that they can easily catch them. The seeds are poisonous raw, but can be eaten when cooked.

IRONWOOD, Ironwood of Borneo

• Ceylon: Billion

Eusideroxylon swageri

An Ironwood is any of a number of trees with extremely hard, heavy wood. This tree from Borneo in Bogor Gardens, Indonesia, is most important for its timber and cabinet wood. There are Casuarinas from Australia called Ironwoods (see "Tropical Trees of Hawaii"—*Hargreaves*— page 28). There is the Lignum Vitae, which is said to be the heaviest of all woods (Page 50, "Tropical Blossoms of the Caribbean"— *Hargreaves*).

25

JACK FRUIT, Jac Fruit, Jak Fruit

· Bali: Nangka · Ceylon: Kos, Pilla-Kai · Fiji: Uto ni Idia · French: Jaquier · Hong Kong: Kwai muk · Samoa: Moe ulu initia · Thailand: Khanoon

Artocarpus heterophyllus

This strange relative of the Breadfruit comes from India and Malaysia. The Indians in Fiji call it "Catara" and use the seeds for curries. The fruits are most unusual. They weigh up to seventy pounds each, and are one of the largest of fruits. They are borne all along the trunk of the tree, and are a very important food in the tropics. The ripe fruit has an unpleasant odor, but the taste outweighs this drawback. The yellowish, soft flaky, sweet pulp is eaten raw, boiled, or fried, and is delicious in curries. The large white seeds are also good roasted, tasting something like chestnuts.

KAPOK, Silk Cotton Tree

Ceylon: Pulungus Gass · Fiji: Vauvaunivavalagi · French: Mapoa, Fromager, Bois Coton, Kapokier · Guam: Atgodon de Manila · Malaysia: Kabu-kabu · Philippines: Doldol, Capoc, Bului · Samoa, Tonga, Niue: Vavae, Lama vavae, Vavae papalangi

Ceiba pentandra

Ceibas are huge, thick grey trunked trees with branches sticking out at right angles, and are often planted in market squares for shade. The leaves have 5 to 9 fingers, and the creamy white 1″ flowers appear just before the leaves. Then the 3″ to 6″ seeds form. It is these capsules that contain the floss called "Kapok" used to stuff pillows, mattresses, life preservers, and upholstery. There are many trees throughout the S. Pacific, Ceylon, etc. In Tonga they are used to mark boundaries.

KOPSIA, White Kopsia

Kopsia singaporensis

This small ornamental evergreen from Singapore has 2″ white flowers with an orange throat that look a little like a Narcissus. Several related species occur in the forests of Java and Malaysia. This one can be found in the Singapore Botanic Gardens.

28

MABOLO, Velvet Apple, Butter Fruit Tree

- Malaysia: Buah Mentaga, B. Saklat
- Philippines, Ceylon: Mabolo (Tag.)

Diospyros discolor

This 40′ tree from the Philippines is also of the Ebony Family. (See page 20.) It has velvety-pink, round 4″ edible fruits that look and taste something like a peach. The flowers are small, creamy-white and faintly fragrant. Seen in the Philippines, Waterfall Gardens, Malaysia, and Hawaii. The lovely big tree above is at Haiku Gardens on Oahu, Hawaii. Fruits may be mistaken for Mangosteens below.

MANGOSTEEN

- Bali: Manggis • Ceylon: Mangus, Mangus Kai
- Thailand: Mang Kud   *Garcinia mangostana*
- Burma: Mingut Thi

This, too, is a small tree with large leathery leaves, native of Malaysia. The round purplish-brown, smooth, but thick skinned fruits the size of an apple are delicious. Many in Bali.

29

MALAY APPLE, Mountain Apple

• Ceylon: Jambu • Fiji: Kavikani, Idia, Kavika • French: Jamelac, Pomme de Tahiti • Guam: Jambu bol, Makupa • Hawaii: 'Ohi'a'ai • Marquesas: Kehika • Niue: i fekakai • Philippines: Jambu bol • Rarotonga: Kehika inana papaa • Samoa: Nonu fi'afi'a • Tahiti: Ahia • Thailand: Chom-poo Kammam • Tonga: Fekika

Eugenia malaccensis

It is said that the Mountain Apple had worked its way across the Pacific before the discovery of America. It was reportedly the only fruit growing here at the coming of the white man. According to Polynesian legend the tree was sacred and temple idols were carved from it. The lovely 50' Malaysian tree has handsome, large shiny green leaves, and tiny cerise blossoms that look like wee shaving brushes. They pop out all over the trunk and limbs and soon the small apple develops.

MANGO, Indian Mango Tree

• French: Mangot, Mangue • Guam:
Mañaga • Indonesia: Pelem Buah
• Malaysia: Pauh • Philippines, Fiji:
Manga • Tahiti, Rarotonga: Vi
Popaʻa • Thailand: Mamoang
• Burma: Thayet   *Mangifera indica*

Originating in India, the Mango is
now planted throughout the tropics,
and is know as the "king of fruits". It
is an evergreen that is not only grown
for its delicious fruit but as a shade
tree, growing to a dense 65′ tree. The
flowers, are tiny pinkish, hairy blossoms forming at the branch tips (see
below). Some people are highly
allergic to the blooms and skins
of the Mango. The maturing fruits
hang prolifically from the tree. When
ripe, their orange pulp is wonderfully
sweet-tasting, juicy, and can be eaten
raw, stewed, frozen or made into
preserves or chutney. They are said
to taste like a peach, apricot, cante-
loupe. The leaves smell like turpentine
when crushed. It is believed that
Mangos have been cultivated by
man for 4,000 years.

MANGROVE, Oriental Mangrove

· Ceylon: Kirilla · Fiji: Dogo, Togo, Dogo-Kana, Nadroga · French: Palituvier · Guam: Mangle Macho · Hawaii: Kukuna-o-ka-la (Rays of the sun) · Philippines: Bacao, Bacuan, Bakawan · Samoa: Toto · Tonga: Tongo

Bruguiera gymnorhiza syn. *B. conjugata*

Native of Malaysia, India and S. China, this is a tree that has many roots growing right in the water. The wood is hard and tough, and the aerial roots were used for bows; roots, fruit and leaves for medicine. The Hawaiians make lovely leis of the many colored (pink, red, yellow) flowers. Many found in Guam. Can be seen along the highway from Nandi to Suva in Fiji. In Tonga, part of the Mangrove was used to put the black coloring in their tapa cloth.

MONKEYPOD TREE, Rain Tree, Saman

· Ceylon: Guango, Inga Saman, Penikaral · Hawaii: 'Ohai · Malaysia: Pukul Lima (5 o'clock) · Samoa: Lopa · Thailand: Chum-sha

Samanea saman

Early travelers reported that the Saman produced rain at night. Legend has it that it "rains from the branches the juice of the cicadas". It is noticeable that the grass is green beneath the tree, no doubt due to the fern-like leaflets closing up at night and the rain falling through—probably why the Malaysians call it a 5 o'clock tree. This lofty patriarchal canopied tree, one of the most symmetrical in the world, is covered with delicate tiny pink-tipped tufts, after which long black pods dangle. These are relished for fodder. The wood is durable and takes a beautiful finish.

33

MONTROUZIERA

Montrouziera, gabriellae baillon, Clusiacae

An interesting tree of the ore ground in New Caledonia is named after the missionary Xaver Montrouzier, who pioneered scientific research and annexation of the island by the French. The tree has leathery leaves and a flower that looks like a Camelia when in bud. As it opens, the yellow stamens that grow in 5 parts, poke out of the center of the pink flower.

NARRA TREE

- Ceylon: Padouk • Fiji: Cibicibi
- Hong Kong: Burmese Rose-
wood • Malaysia: Angsana,
Sena·Philippines: Narra·Burma:
Padauk   *Pterocarpus indicus*

The Narra, the National Tree of
the Philippines, line the streets
of Manila. Its hardwood, either
yellow, red, or white, can be made
into fine furniture and cabinets
that have a roselike fragrance. It
bears numerous yellowish fragrant
flowers followed by small circular
winged pods. The picture above
was made in the National Memorial
in The Philippines.

Also found in Hong Kong Botanic
Gardens; Royal Botanic Gardens,
Ceylon; S. E. Asia, Malaysia and
Fiji,

NONI, Indian Mulberry

· Fiji: Kura · Hawaii: Noni
· Tonga: Nunu, Nonu

Morinda citrifolia

A small evergreen tree with large shiny dark green ovate 8″ leaves is native of Asia, Australia, Guam, Fiji and other islands in the Pacific. It has small white flowers that form in clusters. The fruit is odd, fleshy, edible ovoid warty-looking like a tiny breadfruit that turns from light green to white when ripe. Ancient Polynesians are said to have transported this to Hawaii as they found the bark produces red dye, the root yellow dye, and leaves, fruit and bark are used as a medicine. Used in Fiji, Tonga, and Tahiti. In Tongan myth, Maui was restored to life by the leaves of the "Nonu" placed upon his body. Can be seen in the Medicinal Gardens of the Royal Botanic Gardens of Ceylon.

36

NORFOLK ISLAND PINE

· French: Pin d'Norfolk · Samoa:
Pina · Tonga: Vaini

Araucaria excelsa

This is a perfectly symmetrical evergreen tree discovered by Captain
Cook on Norfolk Island near
Australia in the Pacific Ocean.
Actually, the trees are not pines.
They do not have needles like a pine,
but overlapping scale-like leaves
about one half inch long. These
trees, up to 200' tall, are sometimes
used for masts on ships. They are
also used in landscaping and as
Christmas trees. A closely related
tree is the Cook Pine, *A. columnaris*,
which comes from the Isle of Pines,
New Caledonia. Captain Cook, who
also discovered this tree, said upon
approaching the island, "They had
the appearance of tall pines which
occasioned my giving that name to
the island". It is difficult to dis-
tinguish these two trees apart.

NUTMEG

- Ceylon: Sadhika, Sadhi-kai · Malaysia: Pokok Pala, Pendarah, Pendara-han, Penarah, Penerahan, Buah Pala · Samoa: Atone, Atong-'ula
- Burma: Zalipho Thi   *Myristica fragrans*

This tall 70' to 80' evergreen tree that likes humid conditions is a native of the Moluccas (Spice) Islands. Its small pale yellow flowers that have no petals produce the fruit which has a kernel that is so valuable as the nutmeg of commerce. The ovid 2″ yellow fruit, resembling an apricot, (makes good jellies), about 3 or 6 months after flowering, ripens and splits open disclosing the glossy, dark brown seed. The nutmeg kernel is inside. This seed is surrounded by a net-like scarlet aril which is "Mace", another valued spice—see close up above. 100 nutmegs produce about 3 oz. of dried mace. Mace is called Wasivasi, Poollie in Ceylon. The tree was introduced to Ceylon is 1804. Can be seen in the Royal Botanic Gardens there, one tree is 125 years old. Also grows in Malaysia, Indonesia and the S. Pacific.

38

PAGODA FLOWER TREE,
Yellow Pagoda Flower Tree

Deplanchea speciosa

In New Caledonia, this tree grows wild over the ore ground and in the woods. It is quite a spectacular good sized tree with large green leathery leaves that form on top of the branches. The large clumps of yellow trumpet like flowers have long yellow stamens protruding from the center of the flower. Species are found also in Australia and New Guinea.

PALM TREES

There are over 1500 species of Palms, so only a few of them can be covered here.

OIL PALM

Elaeis guineensis

A stately Palm, native of Tropical Africa, 60′ to 70′ tall that is introducing a coming industry as it produces superior oil (compared to coconut oil) for shortening, vitamin A, soap, margarine, candles, lubricants, and the waste is used for cattle feed. The tree bears when only five or six years old. The many reddish fruits (left picture) are the source of this valuable oil. The outer covering of these date like fruits yields the palm pulp or "pericarp oil" which is used for lubricants and soap, and the kernel supplies the finer "white oil" which is used for margarine and cooking fat. A palm toddy (arrack) is tapped from the tender upper portion of the stem. The Elimina Estates near Kuala Lumpur, Malaysia is one of the first oil estates. It is quite a sight to look down from a plane and see the neat rows and rows of palms planted even on the mountain tops.

PALMS (continued)

TALIPOT PALM, Giant Palm, The Giant of Palms

• Ceylon: Tali gah

Corypha umbraculifera

This native palm of Ceylon is the largest palm tree in the world. It is the National Emblem of Ceylon. It takes 25 to 40 years to grow to maturity and flower, then another two years to produce and ripen fruits, then in about 12 more months, it dies. The Talipot Palm Avenue in the picture at the right was planted in the Royal Botanic Gardens, Peradeniya Ceylon in 1927. Those pictured are now blooming (note spike in tree in the center). The trees grow erect to 80' with a 3' to 4' trunk topped by immense fan-like leaves up to 16' across. The creamy white 25' blossoms appear straight up from the center of the crown of the tree. The fruits are large, hard and marble-like. Small new tree in Bogor Gardens in Indonesia.

SEALING WAX PALM

• Ceylon: Pinang rajah

Cyrlostachys lakka

This clump-forming palm of about 15' comes from Borneo. It has 4' fronds with red leaf sheaths which suggest sealing wax. Avenue at right is in the Botanic Gardens Singapore.

PALMS *(continued)*

IVORY NUT PALM

- Fiji: Niu Soria, Sago, Sogo
- French: Palma de Marfil • Ponape: Och • Samoa: Niu Masoa

Metroxylon amicarum syn. *Coelococcus carolinensis*

This large palm grows in the Caroline Islands. The "nuts" have been exported from the Caroline Islands to Germany where they are called "Steinnuss palme". These seeds are a source of vegetable ivory, as their kernels become very hard when dried, and are manufactured into buttons and other objects. They are interesting to use for arrangements also, as their reddish-brown glossy, scaly, shell's texture makes them look like carved wood. In Samoa, where the light tan "nut" on the left comes from, the trunk of the tree is used for starch. In Fiji they use the fronds for thatch. Also found in Guam, and other S. Pacific Islands.

PALMS (continued)

RATTAN PALM

Daemonorops mollis syn.
Calamus mollis

Rattan is made from the stems of these slender vine-like palms. The stems are long, strong, and flexible, and pretty much of an uniform size. The one pictured above from the Botanic Gardens Singapore, has yards of these stems. Stripped bark yields "peeled cane", which when woven is used for baskets, furniture, tying and binding. Malaysian forests produce excellent rattan.

THORN PALM,
Gendiwung

- Ceylon: Katu-kitul
- Malaysia: Nebong, Nibong

Oncosperma tigillarium

Another clump forming palm has many long black spines that look like big darning needles all along its trunk. The leaves also have spines along their 15' length. The leaf buds are food and the wood of the palm is hard and durable and has many uses. It is a native of S.E. Asia, Malaysia, Philippines; Ceylon, Singapore, and Indonesia.

PAPAYA, Pawpaw, Papay, Papaw, Tree Melon, Mummy Apple

• Ceylon: Papeta, Pepol • Fiji: Maoli, Oleti, Seaki, Weleti, Wi • French: Papaye, Papayer • Hawaii: Mikana, Milikana, Papaia, He'i • Malaysia: Betek, Ketelah • Philippines: Kapayo, Capayo • Samoa: Esi Fafine • Tahiti: Uto, Ita • Thailand: Malakau • Tonga: Lese • Burma: Thimbaw

Carica papaya

There are 45 species of Papaya. It is one of the favorite fruits throughout the tropical world. Bearing fruit when one year old, it is one of the fastest growing trees. Deeply lobed leaves cluster at the top of a hollow trunk under which develop the creamy white fragrant flowers to produce the delicious orange sweet juicy fruit. It contains vitamins A, C, and G, and most parts of the tree contain papain, a digestive enzyme used extensively as a meat tenderizer—tapped from the green fruits and dried.

PAPER BARK TREE, Cajeput Tree, Australian Tea Tree

- Ceylon: Loth-sumbul
- Malaysia: Gelam

Melaleuca leucadendra

Paper Bark Trees are natives of S. E. Asia to Australia. There are many in New Caledonia and Hawaii. Used for reforesting, landscaping, and their unusual annual shed of papery white bark is much valued by Singhalese native doctors in Ceylon, and for torches in Malaysia. Cajeput oil is obtained from these many layers of peeling, spongy bark and also the narrow leaves. It is used as a stimulant and tonic.

45

PARA RUBBER

Hevea brasiliensis

The most valuable rubber yeilding tree is a quick growing 60′ erect tree that is a native of Brazil. It was introduced into tropical Asia in 1877 (Para is a state of N. Brazil). The whitish sap shown dripping from the tree above is the source of most of the world's rubber. The tree has tiny, whitish-yellow panicles of slightly fragrant flowers, and a trilobed fruit capsule which contains inch long brown speckled seeds. Malaysia is one of the worlds greatest producers of natural rubber. Looking down from a plane, there are rows and rows of systematically planted trees. Also grown on Mindanao Island, Philippines; Ceylon, Java, Sumatra, and Thailand.

PERFUME TREE, Ylang-Ylang, Ilang-Ilang

- Ceylon: Wanasapu · Fiji: Makasoi, Makusui, Mokohoi, Mokosoi
- Hawaii: Lanalana · Malaysia: Kenanga, Chenanga, K. Utan, Nyai
  Philippines: Champaca · Samoa: Moso'oi · Thailand: Ka danga · Tonga:
  Mohokoi

Cananga odorata

This 40' native of Burma to Australia is famed as a Perfume Tree because it produces a valuable oil known as "oil of Ilang-ilang" or "Cananga oil". This is distilled from the strongly fragrant droopy 3" to 5" yellow flowers (above), and made into extract as a base for perfume. In the Philippines, Fiji, and many other Pacific Islands, coconut oil is scented with the Ylang-ylang. Samoans make canoes of the wood. Malaysians hollow out the trunk for drums. Rope is made from the bark, flowers and leaves used for medicine, and the wood is also used for tea chests in Ceylon.

47

PILI NUT, Java Almond, Kanari Nut

· Ceylon: Rata-kekuna · New Guinea: Keanee · Malaysia: Buah, Kenari, Pokok Kenari · Philippines: Pili

Canarium commune

A large handsome Malaysian tree with small leaves, small yellowish-white flowers, and pendulous 2″ fruits that turn a dark purple-black when ripe, and have a kernel that tastes somewhat like an almond or walnut—edible raw or roasted. The grey bark contains resin. The tree has an unusual buttressed trunk that is often so circular a section cut out could be used for cart wheels. Canarium Avenue in Bogor Gardens, Java, and the road in the above picture taken on the way to the University of the Philippines, College of Agriculture, Laguna, are both lined with these popular trees. There is also a row of beautiful old Canarium trees with immense base and root structure in the Royal Botanic Gardens, Peradeniya, Ceylon.

48

POINCIANA, Flamboyant, Flame Tree, Fire Tree, Flame of the Forest

· Australia: Crimson Glory · Ceylon: Flamboyante, Giniga, Guli-mohur
· Fiji: Sekoula · Hawaii: 'Ohai-'ula · Philippines, Guam: Arbol de Fuego
· Thailand: Hang nok-yoong · Tonga: Ohai

Delonix regia

One of the most strikingly beautiful trees in the world. Like a huge scarlet umbrella, the tree bursts into dense clusters of red or orange blossoms; one petal of the five petals of each blossom is white—see below. Soon the lovely fern-like foliage appears followed by long brown pods, the seeds of which can be strung into necklaces. There are many beautiful trees in Guam, Hawaii, Cambodia, and Hong Kong where the first trees were planted in 1908 by the Forestry Dept.

49

POISON FISH TREE, Barringtonia, Sea Putat

· Fiji: Vutu Vala, Vutu Rakaraaka · Guam: Puting, Hotu · Malaysia: Sukun, Butong, Butun, Putat Laut · Philippines: Botong · Samoa: Futu · Tahiti: Hutu, Hotu, Tira, 'Utu (heart) · Tonga: Fulu

Barringtonia asiatica

A tall handsome grey barked tree from Asia which is usually one of the first plants to obtain a footing on land, because the heart shaped fruits are buoyant and easily transported by the sea. Fishermen use them for buoys. The ovate 6″ to 18″ leaves form at the branch ends where the fragrant 5″ to 7″ brush-like blooms also cluster. (see left). They open in the evening, and the pink and white blooms can be seen under the tree in the morning. The 4″ fruits contain seeds that can be grated and cast into the water to stun fish—thus the name.

POMELO, Citrus Fruit

· Ceylon: Bambalines, Jambola, Jamblica · Fiji: Moli Kana · Hawaii: Pomelo · Malaysia: Shaddock, Pummelo · Philippines: Suha (Tag.), Pomelo

Citrus grandis syn.
C. maxima, C. decumana (var)

This small 15′ to 30′ nicely shaped tree grows throughout the S. Pacific and S.E. Asia. It is popular among the Chinese who use it for their New Years. The large pear shaped fruits ripen pale green to yellow with pale green, yellow, or reddish-pink pulp. Each segment is covered by strong membranes that can easily be pulled apart. They have a very thick rind. In the Philippines a pink drink mixed with pineapple is made from the juice.

51

POTATO TREE

· Malaysia: Brinjab, Terong

Solanum macranthum

Remarkable because it is one of the few plants of the Potato Family that takes the form of a tree; the Potato Tree is a beautiful flowering tree from South America that was not associated with the Potato Family until the 19th century. It is fast growing to 40' in 2 to 4 years, but it is short lived. Not only do the multicolored 3" flowers of violet, pink, and white make a striking display, but the 10" to 15" dark green leaves also make a handsome contribution of their own. The tree also has many orange berries as big as golf balls. Blooms the year around. Grows in the Philippines, S. Pacific, Malaysia, and was introduced into Ceylon in 1844.

PUA-KENIKENI

· Hawaii, Tahiti: Pua-Keni-
keni · Samoa: Pua-lulu

Fagraea berteriana

Native of the South Pacific
with fragrant 2″ funnel-shap-
ed leathery flowers which are
strung into leis and used to
perfume coconut oil. Pua-
Kenikeni means 10¢ flower
because they used to be sold
for 10¢ apiece. In Tahitian
legend, Pua is sacred to Tane,
God of the Forests, and images of him were made of Pua wood. The flowers open
at sunset; they are first creamy white, then turn orangish-yellow after 5 or 6 days.
The round 1″ fruits are first orange then red.

RAMBUTAN

- Ceylon: Ramtum, Rambutan
- Malaysia, Indonesia, Philippines: Rambutan • Thailand: Ngo-a

Nephelium lappaceum

A handsome spreading Malaysian tree producing large clusters of bright red or orange-yellow fruits which suspend from its branch ends. Each fruit is covered with fleshy soft spines containing a large seed surrounded by white pulp. This pulp is like its relative the Litchi, and is the delicious part of the fruit. They are sold in bunches at little stands along the roads in Kuala Lumpur, Malaysia. Grow in Ceylon, Indonesia, Thailand, Penang, and the Philippines.

54

SIMPOH
Elephant Apple

- Ceylon: Honda Para, Wampara
- Malaysia: Simpoh, Chimpoh • Philippines: Indian Katmon

Dillenia indica

This handsome symmetrical evergreen tree of Ceylon, Malaysia and Asia has 14″ leathery, serate leaves with large showy magnolia-like 8″ white flowers, and 3″ to 5″ fruits that look like big buds or apples. These "apples" never open but fall from the tree still covered by fleshy sepals. Then they are eaten by monkeys, squirrels, and elephants. They can be used for making jellies, curries, and cooling drinks. Can be seen in Botanic Gardens in Hong Kong, Ceylon, Indonesia and Singapore. (the Simpoh Family is divided into two genera: *Dillenia and Wormia.*—see page 63).

55

STAR FRUIT, Carambola

- Ceylon: Kamaranga, Tamarta
- Malaysia: Belimbing Mains, B. Sagi

Averrhoa carambola

A star shaped fruit that grows on a small 20' tree is a Malaysian evergreen that is cultivated for its unusual fruit — sometimes sweet, sometimes sour. The pulp is watery, quite acid but very pleasant tasting. The flowers are tiny red and white clusters that grow on the trunk and branches of the tree (see picture at right). Soon the 2"×5" waxy greenish-yellow to organge five angled fruits develop. Can be used in jelly, preserves, drinks, or raw for salads, making an interesting star shape. The juice removes stains from linens.

CUCUMBER TREE, Bilimbi

- Ceylon: Camias, Bilimbing, Biling, Bilimbikai

Averrhoa bilimbi

This fruit tree is similar to the Carambola above, only instead of having star shaped fruit, the Bilimbi looks like a small 2" to 3" cucumber (see bottom picture). The fruits and flowers are also produced in clusters on the trunk and oldest branches. The fruits are quite sour and juicy. Can be cooked with sugar and eaten in curries, pickles, preserves, jams and drinks. Found in Malaysia, Guam, Hawaii, and Ceylon.

STERCULIA—about 100 species
JAVA OLIVE, Skunk Tree

· Ceylon: Malaiparutti · Fiji: Waci
Waci · Java: Kelumpang, Bangar

Sterculia foetida

This is a deciduous tree from the old
world that has rather unpleasant smelling red, yellow, or purple flowers that
give it a coral color when in bloom.
The fruits resemble cupped hands,
and upon opening reveal smooth black
olive-like seeds. These are edible and
have a purgative effect. Leaves also
medicinal. Another similar tree, *S.
lanceolata*, with greenish or pink
blossoms is most noticeable because
of its fruits. These follicles (center
picture) are bright scarlet pods that
are first green, then yellow, then brilliant red. When opened they form a
perfect 5 petaled rosette with black
pea-sized seeds. Many on Peak Mt.
road in Hong Kong, also in the Waterfall Gardens, Penang; Singapore
Botanic Gardens.

57

STRAWBERRY TREE, Panama Berry, Cherry Tree, Japanese Cherry

· Ceylon: Calabura, Jam-fruit · Guam: Guam Cheri · Malaysia: Buah Cheri · Philippines: Ratiles · Tahiti: Cerise Rose, Cerisier de Panama

Muntingia calabura

This small 30' spreading umbrella-like tree lines the streets of Papeete in Tahiti. The small 1″ white flowers look like strawberry blossoms. They last but one day, then the 5 petals fall off and finally turn into the ½″ pink, then red, smooth berries which are sweet and sought after by children and birds. The tree is very common throughout Guam, Malaysia, Ceylon, and the Philippines. Its leaves are used for tea, the bark for twine.

STRYCHNINE TREE

Ceylon: Goda-kaduru, Kanchurai

Strychinos nux-vomica

Indigenous to Ceylon, India and Burma, this large tree has 2″ orange fruit the pulp of which is eaten by monkeys and birds, but the seeds contain deadly strychnine used for poison arrows, and poisons of both the old and new worlds. Now it is also used as an alternative to digitalis to raise blood pressure. There is a tree in the Royal Botanic Gardens in Ceylon, and Waterfall Gardens, Penang. The above picture is a huge tree behind the administration building at the University of the Philippines. Ceylon and India are the chief sources of supply of the valued medicine.

59

TEAK TREE

- Malaysia: Jati
- Philippines: Tekla
- Burma: Kyun

- Samoa: Pulu
- Thailand: Ton-Sak

Tectona grandis

A large 70' to 150' deciduous tree that lives up to 200 years, and is valued for its strong durable wood that is used for furniture, ships, buildings, bridges, railroad cars, decks, masts and many other things. Teak woodwork has been recorded to last 2000 years. The tree has tiny white flowers borne in three parted terminal panicles up to three feet long. There are many trees in Ceylon and Thailand, where elephants are used to move the heavy timbers. The teak trees in front of the School of Forestry at the University of the Philippines are the oldest in the Philippines. Trees can also be seen at the Botanic Gardens in Singapore, Hawaii, Ceylon.

TEMBUSU

• Malaysia: Tembusi, Tembusu, Temasuk (Borneo), Semesu, Temensu

Fagraea fragrans

This symetrical evergreen tree, which has tiny white-yellow very fragrant flowers and orange berries, is a common avenue and forest tree in Malaysia. It is also seen in Thailand, Sumatra, and was introduced into Ceylon in 1891. Can be seen in the Royal Botanic Gardens Paradeniya in Ceylon, Fiji, Guam, Botanic Gardens Singapore and Penang, where the flying foxes (large bats) come to eat the fruits at night.

61

WATTLE, Acacia

The Wattle is the National Emblem of Australia, and is used in the Australian coat of arms and coinage. There are 500 species of Acacias. They are the most widespread of Australian plants. Many are trees, but some are only shrubs of various sizes. Some of the Australian aboriginal tribes called the Wattles "Mulga, Myall, Boree, Cooba." In Hawaii there is an Acacia called "Koa" that is a native of the Hawaiian forests. Its wood is used for furniture, woodwork, ukuleles and novelties.

62

WORMIA TREE, Simpoh

Another tree of the Simpoh Family. The fruits differ from the *Dillenia* (Elephant-Apple, Page 55) as they split open like a star. Each seed is coated by red pulp which the birds like to eat, (see above.). Another Wormia *Wormia burbidgei* is a smaller sized evergreen tree from Borneo. It has pretty five petaled single yellow flowers and dark green leaves—(see left below.) This is called "Goda para" in Ceylon. Can be see in Botanic Gardens in Ceylon, Indonesia, Hong Kong. Also some species in Australia, New Guinea, Singapore and Malaysia.

63

HONG KONG ORCHID TREE

• Ceylon: Kobonila • Malaysia: Blake's Bauhinia
• Marqueses, Rarotonga: Pine, Fau Kina

Bauhinia blakeana
One of the most beautiful flowering trees is the
official floral emblem of Hong Kong. It was first
discovered there in 1908, now there are many
trees growing all along their roads. The fragrant
rosepurple flower is quite outstanding. Can also
be seen in Guam, Botanic Gardens in Ceylon,
Hawaii and Penang.

Publications by the same authors:

All books in this family are 64 pages each, all
have over 100 full color pictures (some as many
as 130) and all are the same size and format.
Local names in local languages and text are
different to reflect the countries and geographi-
cal areas they cover. Botanical names are
included.

• "TROPICAL BLOSSOMS of the CARIBBEAN"

• "TROPICAL TREES found in the CARIBBEAN, South America, Central America, Mexico"

• "AFRICAN BLOSSOMS" (covers Tropical Africa, South Africa, Madagascar, Mauritius)

• "AFRICAN TREES" (covers same areas as "African Blossoms")

• "HAWAII BLOSSOMS"

• "TROPICAL TREES of HAWAII"

• "TROPICAL BLOSSOMS of the PACIFIC" (covers S.E. Asia, Malaysia, Ceylon and Pacific Ocean countries)

• "TROPICAL TREES of the PACIFIC" (covers same areas as "Tropical Blossoms of the Pacific")

All books can generally be found in book stores and
tourist shops in the particular countries they cover.
Or, books will be mailed postage paid via surface mail
anywhere in the world for $3.85 each in U.S. funds from
the publisher:

ROSS-HARGREAVES
P.O. Box 11897 Lahaina, Hawaii, 96761 U.S.A.
IMPORTANT: If AIR MAIL delivery *outside* the U.S.A.
or territories is desired, add $2.50 U.S. for one copy, plus
$1.00 for each additional copy to cover extra airmail
postage. If AIR MAIL delivery *inside* the U.S. or territories
is desired, add $2.50 for up to two copies, plus 75¢ for
each additional two copies.
When ordering from outside the U.S.A. please send pay-
ment with order in U.S. currency, International Postal
Money Order, Bank Draft, or check on any U.S. bank.

Lithographed in Japan   64